WINDY

by Robin Nelson

first step nonfiction

Lerner Publications Company · Minneapolis

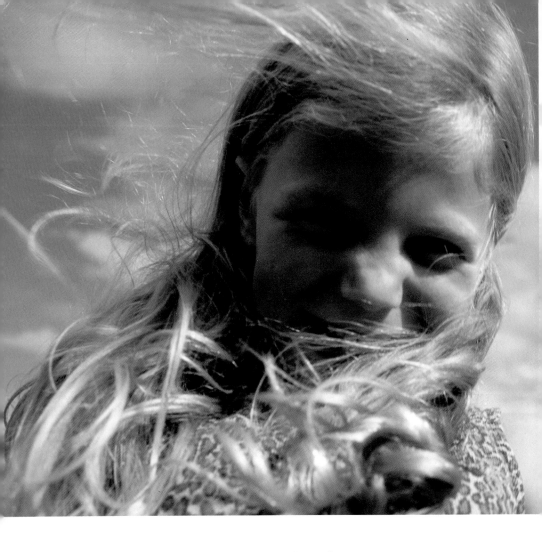

It is windy.

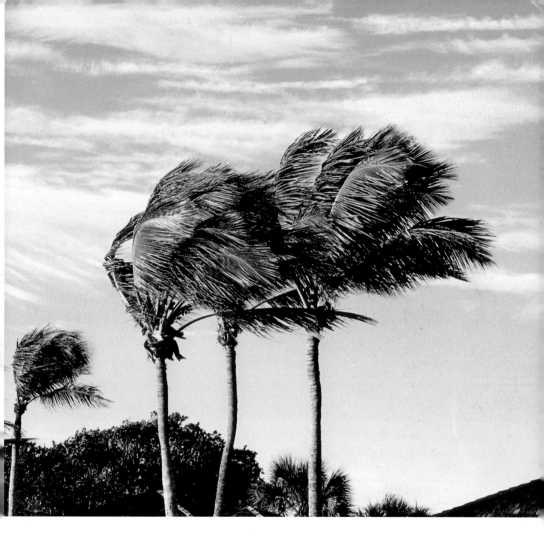

I see trees blowing.

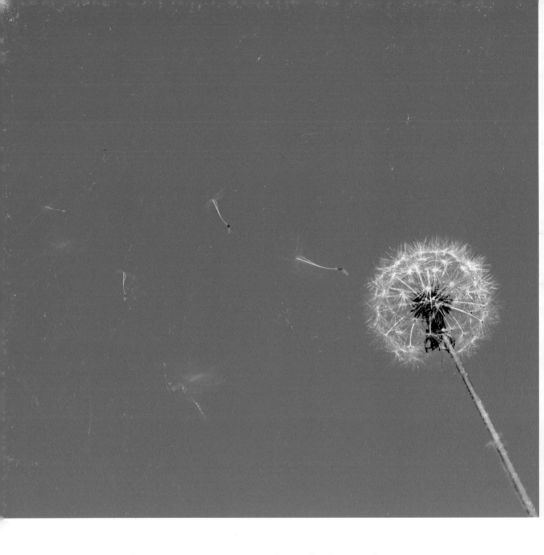

I see seeds blowing.

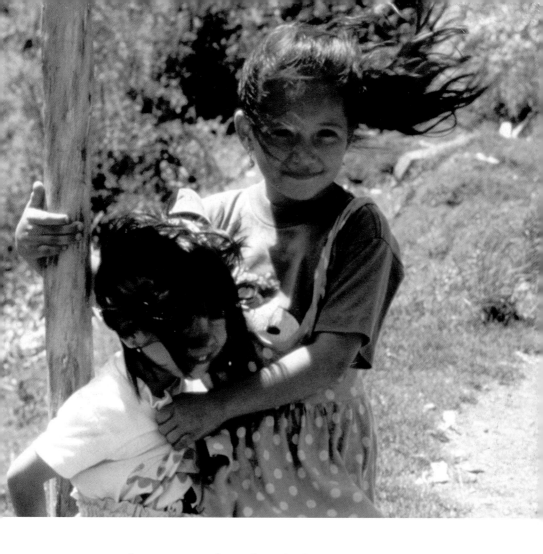

I see hair blowing.

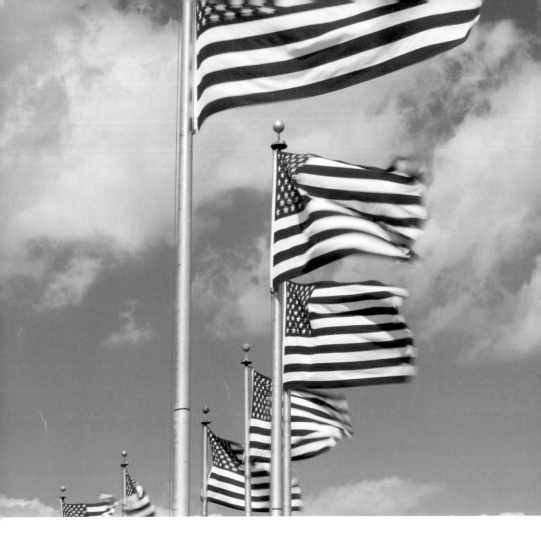

I see flags blowing.

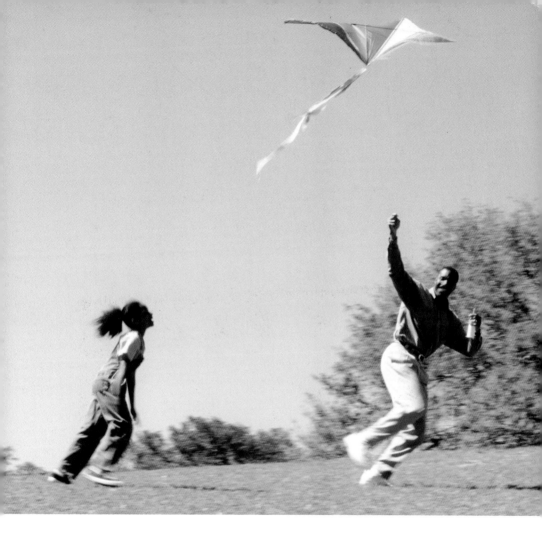

I see kites blowing.

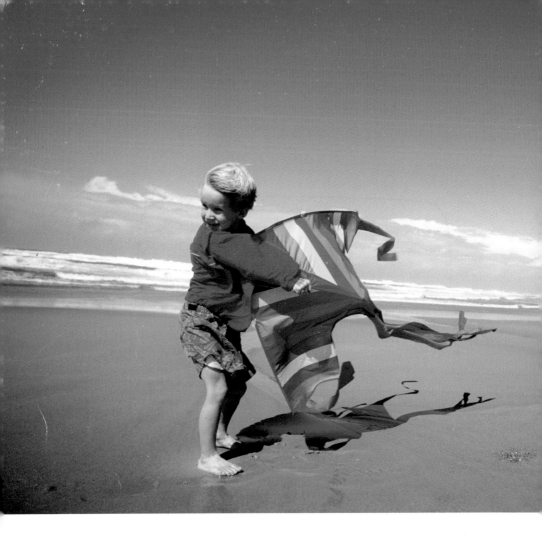

I like windy days!